Originally published under the title *Net mensen* by Lemniscaat b.v. Rotterdam, 2006

Printed in Belgium
First U.S. edition, 2008

Library of Congress Cataloging-in-Publication Data

Schubert, Ingrid.
[Net mensen. English]
Like people / Ingrid and Dieter Schubert.—1st U.S. ed.
p. cm.
Originally published: Netherlands : Lemniscaat b.v. Rotterdam, 2006, under the title, Net mensen.
ISBN 978-1-59078-576-8 (hardcover : alk. paper)
1. Parental behavior in animals—Juvenile literature. I. Schubert, Dieter, ill. II. Title.
QL762.S3213 2008
591.56'3—dc22
2007018429

Lemniscaat
An Imprint of Boyds Mills Press, Inc.
815 Church Street
Honesdale, Pennsylvania 18431

Ingrid and Dieter Schubert

Like People

8

Lemniscaat

Asheville, North Carolina

Many animals in the world have babies—just like people. Furry babies, bald ones, babies with prickles, feathers, or scales. Animal parents teach their babies how to take care of themselves.

Animal parents are everywhere—in the desert, in the snow, even underwater. Animal moms and animal dads hug their children, and when the children play together, the parents always keep them from danger.

It takes two animal parents to have babies. One animal searches for another to love. Sometimes the parents stay together until the children leave the nest. But sometimes the male and female stay together a lifetime. In big families, aunts and uncles help raise the children.

The tasks of raising children are not always divided equally between a mom and dad. Mother sea horse, for instance, doesn't do much after she has laid her eggs in Father's pouch. The eggs hatch a few weeks later. Some fish fathers brood their kids in their mouths. When there is danger, the children swim back in quick!

Father kiwi uses his claws and beak to keep enemies away from his eggs.

But most of the time the mother builds the nest and raises the children—sometimes by herself, sometimes with a group of other mothers.

Many animal babies are born naked and blind. They don't look like their parents yet, and they are completely helpless. Kangaroo and platypus babies are not much bigger than a bean! The tiniest babies live the first period of their lives in a pouch or on their mothers' bellies. They drink their mothers' milk until they are big enough to walk.

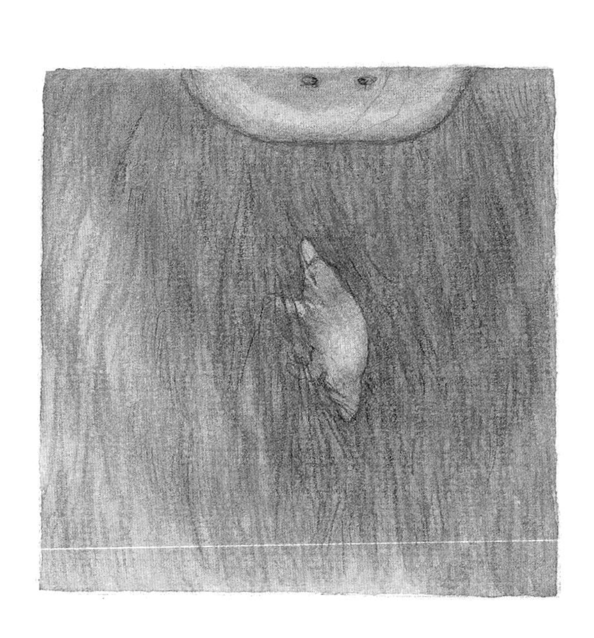

At birth, the babies of elephants, rhinoceroses, pigs, tapirs, ducks, giraffes, and guinea pigs usually look like their parents—only a bit smaller. They follow their mothers right after they are born.

Some animals start their lives drinking their mothers' milk. Mom is always around, and nursing is warm and cozy.

Other babies get food that has been pre-chewed by their parents. Or Mom and Dad bring snacks all day long. When you are an animal, it is better not to have too many siblings because the strongest baby always gets the most food. The weakest has to wait for leftovers.

After dinner, some babies get a bath. Not all babies are amused by bath time. Some females that live in herds take turns washing all the babies. That way the other mothers can take a nice quiet bath every once in a while.

If a baby that is part of a herd loses its mother, it will often be adopted by another mother, so it will still grow up in safety.

Babies know a lot when they are born, but what they don't already know they learn by watching their parents closely, imitating them, and practicing a lot. That's how they learn to hunt, flee, defend, and attack.

They also learn how to protect themselves against sun and rain and how to build a nice home. Of course, they learn all the best tricks for getting the tastiest snacks.

Young animals play together and learn how to get along, just like people. Sometimes young males tease and challenge the older animals. The teasing is not always appreciated and sometimes leads to small fights.

Young females find their place among the older females. When there are too many males, the younger males leave the group. Sometimes they don't want to leave and have to be chased away.

Then the story starts all over again. The young animals try to find another animal they like. A male spots a cute female, and, imitating his father, pretends to be really tough. The lion shows his mane, the peacock shows his feathers, and the walrus his teeth …

while other animals rely on their voices.

Because animals like a beautiful love song—just like people.